Bergbuchlein, The Little Book on Ores

The First Mining Book Ever Printed

A translation from the German of the
Bergbuchlein

a sixteenth-century book on mining geology
attributed to Ulrich von Kalbe

translated by
Anneliese Grunhaldt Sisco

with technical annotations and historical notes by
Dr. Cyril Stanley Smith

compiled by
Dr. Mary Ross

Oxshott Press

2

Published by: Oxshott Press, England

ISBN 978-0-9568322-3-8

ACKNOWLEDGEMENTS

I am pleased to acknowledge the American Institute of Mining, Metallurgical, and Petroleum Engineers, Inc. (AIME) for translating the sixteenth-century text *Bergbuchlein,* "The Little Book on Ores," the first text ever written on mining. The care and expertise that went into the translation, performed by Aneliese Grunhaldt Sisco with technical annotations and historical notes by Dr. Cyril Stanley Smith, was extraordinary; no detail was overlooked in creating an accurate account of this early German work to make it accessible to an English-speaking audience. Funded by a grant from the Seeley W. Mudd Memorial Fund, the AIME published their translation of this historic work in 1949, along with the translation of a slightly later German text, *Probirbuchlein* ("the Little Book on Assaying") in a volume that combined the two works. This combined translation was titled *Bergwerk Und Probirbuchlein.*

I am further indebted to AIME and their Member Society, the Society for Mining, Metallurgy, and Exploration (SME), for generously granting permission to reprint their translation of the *Bergbuchlein* in this current volume.

As noted in AIME's *Bergwerk Und Probirbuchlein,* the illustrations presented to accompany the text of *Bergbuchlein* have been reproduced from various editions of the book found in libraries and private collections around the world. These include material from the following sources:

4

The Bibliotheque Nationale of Paris for an early but
undated edition of *Bergbuchlein* (A),

The Osler Library of McGill University, Montreal, for
a 1518 edition published in Worms, Saxony
(B) and a 1535 combined edition published in
Frankfurt (G),

Mr. Herbert Hoover, Jr., for a 1527 edition published in
Erfurt (C) and a 1535 combined edition
published in Frankfurt (E),

The Parsons Collection of the Reference Department of
the New York Public Library for an early but
undated edition (D) and

The Library of the Bergakademie Freiberg, for a 1534
combined edition published in Augsburg (F).

As can be seen from the number of copies examined for this translation, the AIME's dedication in documenting this historic work and presenting it to an English audience was exemplary. I am indebted to their unstinting attention to excellence in creating the translation, and to the AIME and the SME for allowing me to reprint it here in this modest volume.

I would also like to thank Dr. Andrew Michael from the U.S. Geological Survey who suggested creating this edition. And a very special thanks to Conrad Richard Kolbe whose love of history and scholarly curiosity inspired it.

- M. R.

June 2013

FOREWORD

In 1949 the American Institute of Mining and Metallurgical Engineers (AIME) published a small volume titled *Bergwerk Und Probirbuchlein* which was an English translation of two German sixteenth-century works. These included the *Bergbuchlein,* "the Little Book on Ores," the first printed work in the field of mining, and the slightly later *Probirbuchlein,* "the Little Book on Assaying."

As the first book on mining ever written, *Bergbuchlein* has particular historic significance. For this reason, it is being presented here in this volume. The translation of *Probirbuchlein* is not included here. For that, the reader is referred to the AIME's 1949 edition of *Bergwerk Und Probirbuchlein,* which is currently out of print but can sometimes be found through out-of-print sources.

Bergbuchlein was originally written anonymously sometime after 1500. The first dated edition of the book is 1518, although there is an earlier undated version that some scholars believe might have been written between 1505 - 1510. The text was later attributed to "Calbus of Freiberg" who is now considered to be Ulrich von Kalbe, a physician who lived in Freiberg in the Saxony region of Germany between 1454 and 1523. A further discussion of the history of the work is provided later in this volume.

Probirbuchlein, written shortly after 1520, was also published anonymously. Unfortunately, no author has been attributed to that work.

For many mining professionals, it is Georgius Agricola who is considered the father of mining, based on his publication of his early mining text, *De re Metallica.* This massive volume was published in 1556, 32 years after von Kalbe's death. Few are aware that Agricola used material from earlier sources, notably the *Bergbuchlein* as well as *Probirbuchlein* in his massive treatise. While Agricola's work was considerably more comprehensive than von Kalbe's earlier work, the historic nature of the *Bergbuchlein* as the first mining text ever published is undeniable.

It is of note that Ulrich von Kalbe's role in authoring the first mining text is poorly known, even in mining circles. In von Kalbe's hometown of Freiberg, Germany, which hosts the Bergakademie Freiberg, the oldest mining university in the world, the university's library is named after Georgius Agricola, who was a scholar from a different city. It was Agricola himself who credited certain excerpts from the *Bergbuchlein* in his text and unmasked von Kalbe as the author of that anonymously published first book on mining. While the modest *Bergbuchlein* with its references to alchemy and astrology can not compare to Agricola's later but more comprehensive work, *Bergbuchlein's* historical significance as the oldest mining text remains.

Regarding the translation from the German to English, AIME employed a collaborative approach between Anneliese Sisco, a linguist with a scholarly understanding of the German language, and Dr. Cyril Smith, an expert on minerals and metals. This was done to provide accuracy not only in terms of the overall language but also for technical understanding as certain terms have shifted meaning over the years. This combined approach delivered a translation that is not only accessible but makes fascinating reading even 500 years after the text was first written. The woodcut illustrations, reproduced faithfully from the sixteenth-century originals, are lively and energetic, and provide an excellent accompaniment to the written text.

Written as a dialogue between Daniel, a mining expert, and young Knappius, a student of mining, *Bergbushlein's* text summarizes the known mining knowledge of the day. Daniel reminds his young student to be mindful not only of the profits of mining but to embrace the "art" of the field and appreciate the wonders of nature and the "mineral Power."

In the spirit of historic preservation, and in appreciation of Ulrich von Kalbe, a pioneering soul who did something no one had ever done before him, I am pleased to present this translation of his work once again in order to make *Bergbuchlein* available to a modern audience.

M.R.

June 2013

INTRODUCTION

The *Bergbuchlein* occupies a unique position in the literature of mining as it is the first printed book on any aspect of mining. The first printed book on any subject has an allure that may not be entirely warranted by the strict historical significance of its content. Among the books of the first century of printing there are very few that contain information that was new at the time of publication. Much of what they recorded had been previously circulated in manuscript form. Printed books are nevertheless of great value because they mark the time when the author's words, free from errors -- or at least with the same errors-- became available to large numbers of readers, who thus could easily and cheaply acquire a common background for their future work. Moreover, though there is occasionally some doubt as to the exact year of publication (as in the case of this work), it is generally easier to date a printed work than a manuscript, and thus to provide a fixed coordinate in history.

The *Bergbuchlein*, "The Little Book on Ores," is the fist printed work in the field of mining. It is an introduction to mining geology, which means that it was not addressed to the practicing expert but was intended to rouse the interest of beginners in the various aspects of a future vocation. It touches on the theories of the generation of ores, introduces and defines some of the most frequently encountered technical terms of the profession, and indicates what knowledge and tools are required for successful prospecting and mining. It also describes the ores of the seven most important metals. For those parts that deal with theory, the author accepted uncritically the traditional teachings of the alchemists and the astrologers; his hints on where to find promising ores are based partly on superstition; but there is enough factual, practical information, for example, on veins and their differences, on the occurrence of a metal in different ores, and on the association of a specific ore with others or with certain gangue materials, so that even a scholar like Agricola used the *Bergbuchlein* as a reference book.

After its first appearance, which was probably not much later than 1500, it was reprinted at frequent intervals. Not only do the number of reissues but the diversity of local of publication attests to the popularity of this work.

There was naturally a demand for information on some phrases of metallurgy other than assaying. The *Bergbuchlein* ends with the promise of a description of smelting. As far as is known, this promise was not kept. The publisher Christian Egenolph tried to supply the demand by combining the *Bergbuchlein* with parts of the *Probirbuchlein*, "The Little Book on Assaying," which was written around 1520 as the first book printed on metallurgy. The publisher particularly used those sections dealing with assay furnace and cupels and added some miscellaneous matter on dissolving metals, on polishing gems, and on metal poisoning. This combined volume, first printed in 1533, was called *Bergwerck und Probirbuchlein.*

Most of the miscellaneous matter in the original omnibus volume is also found in the *Kunstbuchlein,* which was published first under that title in 1535 by Steyner in Augsburg, though earlier books of similar content are known. The subject matter and bibliography of its many varied editions have been well discussed by Ferguson*. This series is of interest from a metallurgical point of view, for it contains, mixed up with various cosmetic and medicinal nostrums, a number of metallurgical recipes of importance-- advice on such operations as hardening steel, soldering and brazing, amalgamation, gilding, making metal powders, and casting medallions, as well as that interesting antecedent to the flotation process, the

---

*John Ferguson: Some Early Treatises on Technological Chemistry, *Proc. Pbil. Soc.* Glasgow, v. 19, 1887, pp. 116-159; v. 43, 1911, pp. 232-258; v. 44, 1912, pp. 149-189.

purification of lapis lazuli by mixing with a greasy paste and kneading under water.

In his Berg-, Probir- und Kunstbuchlein (Munich 1926), Ernst Darmstaedter provides a comprehensive bibliography and discussion of these three somewhat diverse works. The joint consideration is justified, for none of them was meant to be a learned treatise of its subject but all furnished useful technical information cheaply, probably to the annoyance of some master artisans who must have objected to the revelation of their Mysteries. The closest modern analogues, indeed their lineal descendants, are the books of household recipes, the mechanic's, aids, the chemical formularies, and the prospector's handbooks that were so common-- and so badly edited and printed -- in the last century and are to this day occasionally issued, doubtless to confuse the future bibliographer and to give him a strange idea of the slowness of dissemination of present-day scientific knowledge.

The present translation of *Bergbuchlein* was made from the 1518 edition but other editions were carefully examined and significant differences in earlier versions are indicated.

From notes by Anneliese G. Sisco and Cyril S. Smith,
October, 1948

CONTENTS:

12

A well-planned, useful little
book on how to prospect for and find the
ores of the different metals, with illustrations of the lay of the terrain and
an appendix of mining terms,
which will prove most useful to young miners.

==

Bergbüchlein

DANIEL, THE MINING EXPERT, TO
YOUNG KNAPPIUS

COMPLYING with your frequently expressed wish and your persistent request, I have decided to prepare a brief little book on metallic ores, based on the books of the ancient philosophers and on the experience of practicing miners, that will instruct you how to judge from promising indications which rock formations, which veins, stringers, or mineral matter are likely to bring forth metallic ores so that it would be promising and gainful to work them. It explains according to the best of my ability, in separate chapters, as much as you need to know for your chosen profession or business, about such things as the favorable characteristics of the terrain, the strike, and dip* and outcrop of veins in general, and also about each metal in particular.

Young Knappius:† Then it will be explained to me logically in this little book and I shall be given a reasonable understanding which mines can be worked gainfully so that my investment will not be wasted but will show a profit.

Daniel: Everybody should, of course, use his intelligence and should educate himself well and diligently to become able to understand, [2]‡ as far as that is possible, from what matter, by what means, and where metals are generated by Nature; and as a mere side issue he should not spurn the profit that ensues therefrom. But if his aim is solely and predominantly profit and gain rather

---

*The German words *Streichen* and *Fallen* have the same technical meaning as "strike" and "dip." But in the mind of the author the words do not have the same accurate meaning as in the scientific literature of today. T.F.W.B.

†*Knappe* means miner.

‡The numbers in brackets represent the page numbers in the original.

than the desire to know about the wondrous influences that Nature works under the earth by means of the mineral Power, it would cheapen and condemn this little book and the art. If you really value gain more than the art, you will have to do without either art or profit. And furthermore, consider this well: the general talk that follows on such things as the position, the strike, footwall, and other characteristics of veins often has to be applied with great adroitness to this or that special case.

Knappius: To become expert in that I shall need practice.

Daniel: If you understand the items in this little book dealing with the quartering of the earth, and if you also practice the art of mining, you will gain great insight into the working of Nature under the earth.

Knappius: Since you are going to discuss the quartering of the earth and the division of a mine, I want to ask this: In what direction or face in the shaft or at what corner of the drift would my or Lamprecht's share be located?—so that I can see in the pit what profit I can expect.

[2v] *Daniel:* Your ignorance about mining has driven me to undertake this job. But do you actually think that a share is some specially defined location in a mountain? If that were so, equal assessments* and contributions would be applied to very unequal values. In reality, a share is the 128th part of all that belongs to a mine. The mine assets are conveniently divided into 4 blocks (each of which may be divided into 8 shares, 16 shares, 32 shares, or 64 half-shares) and a maximum of 128 shares; and the share is divided into half-shares, quarter-shares, and halves of quarter-shares. Thus the division is made by means of a number called *pariter partite* [the common denominator]. To increase a person's holdings you can add only enough equal fractions to it to reach unity, but

---

*This refers to the coöperative working of mines, or parts of mines, by the *Gewerkschaft*. The shares are *Kuxe*, and the block of 32 shares is a *Schicht*.

you can continue indefinitely to divide this fraction into equal parts to decrease the magnitude of the holding, and only this will tell you into how many shares a mine is really divided.

Do not mind if this little book uses simple words and unpolished phrases. They convey something useful, which you should esteem higher than the smoothness of words.

Half the shift is up* and lest we prolong our shift unnecessarily I shall add only that, for a better understanding of the origin and source of metallic ores, this little book on the birth of ores† and minerals is divided into ten chapters.

---

[3]   The First Chapter

About the Origin of Ores in General, whether Silver, Gold, Tin, Copper, Iron, or Lead Ores.

IN which they are all introduced under the common name: metallic ores. It should be realized that for ores to grow or to be born requires an agent to exert an influence, and a passive thing or matter that is qualified to be influenced. In the words of the naturalists, the common maker of ore and all other things that are born is Heaven with its movement, radiance, and influence. The influence of Heaven is diversified by the movement of the firmament and the countermovement of the seven planets. In this way each metallic ore receives an influence from its own particular planet, specifically assigned to it because of the characteristics of the planet and the ore, and also because of their conformity in warmth or frigidity, moisture or dryness. Thus, gold is made by the Sun or his influence, silver by the Moon, tin by Jupiter,

---

*The German equivalent for "the shift is up" or "has ascended" has become an adage meaning "enough words about that."
†*ettlich* misprint for *ertlich*.

Bergbüchlein

copper by Venus, iron by Mars, lead by Saturn, and quicksilver by Mercury. That is why Hermes* and other learned men often call the metals by these names, that is, they call gold sun, in Latin *sol*, and silver moon, in Latin *luna*. This is brought out more clearly in the separate chapters on each metal, [3*v*] so that this short statement about the common maker of all metals and ores shall suffice here.

But the passive thing, or the common matter of all metals, is, according to the opinion of the philosophers, sulphur and quicksilver, which, through the movement and influence of Heaven, must be joined and hardened into a metallic body or an ore. Some think that through the movement and influence of Heaven vapors or fumes (called *exhalationes minerales*) of sulphur and quicksilver are pulled up from the depth of the earth, which, when ascending through fissures and fractures [which become the veins and stringers]† are united by the influence of the planets and are made into ores. But there are others who do not believe that metals are made from quicksilver because metallic ores occur in many locations where no quicksilver is found. They assume, instead of quicksilver, a moist, cold, muddy matter, without any sulphur, that exudes from the earth as if it were its sweat, and think that all metals are made by its commingling with sulphur.

But never mind; if you understand and interpret them correctly, both theories are right; that is, ore or metal is made of the moisture of the earth, called matter of the first order, and of vapors and fumes, called matter of the second order, both of which shall here be called quicksilver. Thus, in the mingling or union of quicksilver and

*Hermes Trismegistus, the fabled author of the *Hermetic Books* on alchemy and other subjects.

†The German here reads *in gängen und klüften*; the bracketed interpolation seems necessary because the German words, being used for the openings in the rock (fissures and fractures) as well as for veins of different sizes (veins and stringers), mean the former here but introduce the names for the latter.

sulphur in ore, sulphur acts as the male seed and [4] quicksilver as the female seed in the birth or conception of a child. That is the story of sulphur as a special, qualified maker of ores or metals.

The Second Chapter

About the Qualifications of the Terrain in General.

ALTHOUGH an influence from Heaven and favorable characteristics of matter are essential in the making of every ore or metal, that alone is not enough. If ores are to be born readily, it also requires favorable characteristics of the natural vessels in which the ore is made. These are the veins, such as steep veins,* sloping veins, branch veins, cross veins, or whatever they may be called regionally. It also requires channels or means of access through which the mineral- or ore-creating Power can easily enter these natural vessels. These are provided by the stringers such as hanging-wall, intersecting, sloping, and cross stringers, or by approaching flat veins which are given different names regionally. It also requires a favorable lay of the terrain which the veins and stringers traverse. Ore deposits may be oriented in any direction,

---

*The translation of the names of veins and stringers is based mainly on definitions in Delius, Veith, and the *Bergwercks-Lexicon* (see p. 63). Thus:

a *stebender Gang* or *steinender Gang*—the latter undoubtedly a misprint—translated "steep vein," is perpendicular or nearly so, with a dip of 80 to 90 degrees;

a *flacher Gang*, translated "sloping vein," is not a flat vein but dips from 20 to 50 degrees;

a *schwebender Gang*, translated "horizontal vein" or "flat vein," is nearly horizontal. The maximum dip is 20 degrees, in some regions only 5.5 degrees. This expression is sometimes interchangeable with *Flöz* or *Flözwerk*; but the latter may also be a bedded deposit, i.e., running parallel to the rock strata.

A stringer is a *Kluft*, or a *Trum* in more modern German.

sometimes toward morning,* sometimes toward noon, sometimes toward evening, and sometimes toward midnight in the slope of a mountain. But the side of a mountain, or a location, facing noon is more likely than any other one [4v] to bear rich ore, especially if there is a gently sloping stretch of lower country lying to the south. Such characteristics of a terrain are the most promising for the occurrence of workable ore. The following figure will illustrate this.

[Fig. 1]

For a better understanding of what has been said about the quarters of the earth—and of what is to follow —it should be noted that the whole earth is divided into 24 parts. This is done by means of a circle called horizon,

*In later chapters this has been translated east, south, west, and north, respectively.

The Terrain

which divides heaven into an upper and a lower part,
[5] so that what you see of heaven encloses the earth.
This circle is divided first into four parts by means of
two lines that cross each other at equal angles or corners.
These are called sunrise or morning, noon, sunset or
evening, and midnight. After that, each part is further
divided into six sections. Morning is given the number 6;
the numbers 7, 8, 9, 10, and 11 are given to the next
sections before noon. Noon is 12, and the sections after
noon are numbered 1, 2, 3, 4, and 5. Evening is 6, and
the numbers 7, 8, 9, 10, and 11 are given to the sections
following evening. Midnight is 12, and the numbers 1,
2, 3, 4, and 5 are given to the sections after midnight,
just as time is divided into two halves. The following
figure will serve to show this better.

[Fig. 2]

[5v] # The Third Chapter

On the Strike and the Outcrop of Veins and Stringers.

THE strike of a vein is the direction in which it pro-
ceeds lengthwise in the rock of the terrain. Sometimes
a strike is from east to west, sometimes from west to east.
A vein strikes from east to west if the rock of its hanging
wall [6] (with its little earth-filled joints*) dips west-
ward; and vice versa, a vein strikes from west to east if
its rock dips eastward, as is shown in the accompanying

[Fig. 3]

Schmerklüftlein. This may mean any fracture filled with sticky earth.
But according to the pictures (Fig. 1 and Figs. 3 to 6) the author here more
specifically may have thought of feather joints that sometimes will form
along the edges of a vein. T.F.W.B.

Veins

figures. There is a further differentiation according to the slope of the terrain.* The first figure [Fig. 3] illustrates a south slope, the second figure [Fig. 4] illustrates a north slope.

[6v] Also, some veins strike from south to north and others from north to south. The distinction is made by

[Fig. 4]

*It seems as if the original of this statement, which may be interpreted in several ways, and the following sentence caused as much trouble to the illustrators of the *Bergbüchlein* as they did to its translator. In the 1518 edition and all those reprinted from it, Figs. 3 to 6 are obviously meant to illustrate an east-west, a west-east, a south-north, and a north-south strike, in that sequence; and the illustrations reflect the artist's belief that the difference between an east-west and a west-east strike, for instance, depends somehow on their occurrence in a south and a north slope, respectively. Apart from the fact that, as drawn here, both Fig. 3 and Fig. 4 show a south slope, viewed from the east in the first instance and from the west in the second, this is clearly a misinterpretation of the author's meaning—who made the distinction by the orientation of the country rock and used the slope of the mountain only for further description (see p. 35)—and of

Bergbüchlein

the dip of the rock, as explained above, and again a further differentiation is made by the slope of the terrain as
is illustrated here [Figs. 5 and 6].

Der Morgen

Der Abend

[Fig. 5]

the illustrations in the earlier edition from which the present ones were copied.

In the illustrations of the undated edition, it is made clear by the shading
of the veins that in each figure the two veins are meant to represent opposite
strikes. Thus, Fig. 3 was meant to show an east-west and a west-east strike
in a south slope and Fig. 4 an east-west and a west-east strike in a north
slope. A copy of these two figures as they appear in the undated edition
is given on p. 66. It shows incidentally that the repetition of *Morgen* and
Abend in the corresponding figures of the present edition was caused by
the crowding in the earlier edition, which was confusing to the copyist.
Other errors, which increase in number with every reprinting, are to be as-
cribed to the fact that the legends around the woodcuts were not part of
the blocks.

Figure 6, with the veins shaded to indicate opposite directions, was meant
to illustrate a south-north and a north-south strike in an east slope. In
Fig. 5, the directions are hopelessly confused, and they already appear that
way in the earlier edition. This figure should illustrate a south-north and a
north-south strike in a west slope; i.e., the legends should read as in Fig. 6.

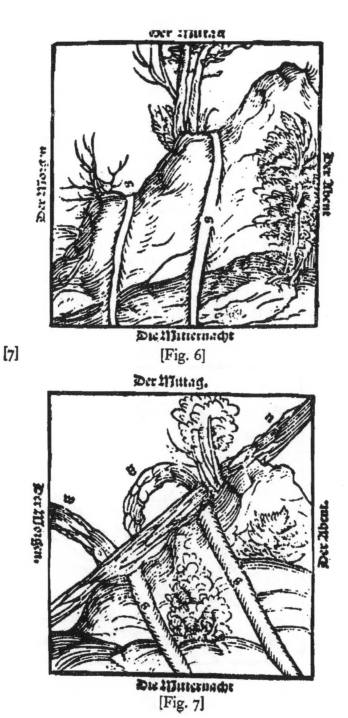

Der Mittag

Der Morgen

Der Abent

Die Mitternacht

[7]

[Fig. 6]

Der Mittag.

Der Morgen.

Der Abent

Die Mitternacht

[Fig. 7]

Bergbüchlein

[7v] Some veins strike from midway between east and south toward midway between west and north, and still others from midway between west and north toward midway between east and south.

[8] Also, some veins strike from midway between south and west toward midway between east and north, and some again from midway between east and north toward midway between south and west. Again, this is differentiated according to the slope of the terrain and is illustrated in the following figure.

Some veins strike between any other point of the compass and the one opposite it in either one of two directions. [8v] Thus, veins that are straight and simple may run in any one of the 24 possible different directions shown clearly earlier in the figure illustrating the quartering of the

Der Mittag

Die Mitternacht.

[Fig. 8]

Veins

earth. However, there are other veins whose strike is not straight and simple but which are curved like a half-circle or disturbed in their original direction by the junction of another ore body. Some may then run from east to south and others from south to west or in some other direction. These veins that are irregular [9] in their strike are also irregular in their ore shoots as is outlined in other chapters.

And then there are some veins that run in gently rolling fields, which explains the term "field mining";* and others that run through a depression or valley, from east to west or vice versa, or from south to north or vice versa, or in any of the previously described directions. But

[Fig. 9]

*Apparently the author was puzzled by the use of the word *feldbau* as a synonym for *grubenbau*. The English language makes similar use of "field" in such expressions as coal field, diamond field, etc.

26

Bergbüchlein

enough of the strike of veins. Let us now consider the
hanging wall and the footwall.

[9v] Every vein has a hanging wall and a footwall. The
hanging wall of a vein is the roof over the vein, which is
touched by the back of the vein. Its footwall is the rock
on which it rests. But there are some veins whose dip
is so steep that their hanging wall can be distinguished
from their footwall only by means of branching hanging-
wall stringers, which may give an indication; [and these
hanging walls and footwalls may also be oriented in]*
any of the previously described directions. For illustration consult the preceding [sic] figure. But enough of

Der Morgen

Der Abent

[Fig. 10]

*The interpolated line is taken from the undated edition. Figure 10 of
that edition (A) shows the hanging-wall stringers, the importance of which
escaped the copyist.

Veins

hanging walls and footwalls. The following deals with the outcrops of veins.

To every vein belong two outcrops. One is where it emerges into daylight along the whole length of the vein and is called the outcrop of the whole vein. The other one is in the direction contrary or opposite to the strike of the vein, as judged from the orientation of its rock, and is called the rock outcrop. Thus, every vein whose strike is from east to west has a rock outcrop in the east; and vice versa, if the strike is from the west, the rock crops out in the west. This is true also for the other quarters of the earth so that, depending on how the vein strikes, the rock outcrop may be in every possible direction, as you can see easily from the preceding figures. So much for the outcrops belonging to a vein.

For a better understanding of what has been said so far [10] about the quarters of the earth and about the strike of veins, it should be noted that a compass must be divided into 24 parts by a special circle.* First by two lines that intersect, forming equal angles or corners, into four parts. One of the lines shall run from 12 of the compass over the magnet, or over the center of the little iron needle (that has received its force from the magnet), directly to the center between 5 or 4 in the forenoon [and 7 or 8 in the afternoon], depending on how the compass is made. The other line, as was said previously, shall intersect the line just described at equal angles. And as, depending on whether a country is situated in the south or in the north, the pole rises higher and higher over the circle of the earth (called the horizon), so runs the afore-

*The following description is quite unintelligible. Comparison with the text of the undated edition showed this to be due, in part, to omissions in the present edition. In the translation, these omissions are interpolated in brackets. In addition, there seem to be a number of misprints, which occur in all of the editions known. These have been preserved in the translation. The illustration of the compass of the undated edition is reproduced on p. 65. This is clearly a compass with two circles to allow for magnetic declination. One of the circles is aligned with the needle, while the other gives true directions immediately.

Bergbüchlein

mentioned intersecting or crossline from different points of the compass. And furthermore, depending on whether the circumference is drawn inside or outside of the circle marking the hours, it differs more and more. Sometimes it runs from before 8 of the morning to before 4 of the evening, sometimes straight from 8 to 4 [sometimes from after 8 to after 4]. Watch, concerning the lines of the compass that mark the hours of the day, that the difference between the markings of the cross and the markings of the hours, which deviate a little, does not confuse you. But never mind; here, in the blessed country of Meissen, this crossline on the [10v] compass generally runs from just before 8, where its line touches the lower circle, to just that much before 4. The chronographers are well aware of this; and, therefore, the line at 4 after noon must always be marked 6 of the morning or sunrise,

[Fig. 11]

Veins

because this line of the compass always means morning, and the line between 4 or 5 before noon and 7 or 8 after noon must always be marked 12 noon because this line on the compass always points to noon. And [11] the line at 8 before noon must be marked 6 of the evening because this line always points to evening; and the line at 12 of the compass is to be marked 12 midnight, because this line always means midnight. Since each part is divided into six sections as before, the whole world is divided into 24 parts, as shown in the preceding figure.

Thus you can obtain definite knowledge of the orientation and of the strike, dip, and outcrop of veins, if a compass so divided is held over a vein.

The Following Deals with Stringers.

WHAT has been said about the strike of veins also applies to the strike, dip, and outcrop of stringers, because some stringers strike or pitch from east to west, some from south to north or the opposite, and some in still other directions. Among the stringers are hanging-wall stringers, intersecting stringers, and cross stringers, or whatever they may be called by mining men. Some of these cause or contribute to the enrichment of a vein and produce a rich ore. Others deprive and rob the vein of the mineral Influence or ore-producing Power, which is the reason why sometimes a distinct shimmering* is ob-

*The German word is *Witterung*, which according to Veith's *Deutsches Bergwörterbuch*, is sometimes called *Bergfeuer*. In this sense, *Witterung* obviously refers to an atmospheric phenomenon reputedly observed near the outcrop of a rich vein. It was believed to be a manifestation of "mineral exhalations." If a thieving stringer would give rise to this phenomenon, a prospector would be misled to think that he had discovered a rich vein.

But *Witterung* is also a mineral-bearing rock decomposed by the exhalations whose function it was not only to generate ores but, after having done so, to destroy them again. *Witterung* in this sense is used frequently later in this book and is then translated "alteration product."

F. D. Adams discusses *Witterung* at some length in his *Birth and Development of the Geological Sciences* and states that German miners are no longer

Bergbüchlein

served far from a vein, so that many a mining man has
become confused in his judgment. Which of these stringers cause enrichment and which cause impoverishment
of veins will be shown in the next chapter.

The Fourth Chapter

On Silver Ore and Its Veins.

IF one were to follow the order in which Nature gen-
erates things, the imperfect metal should rightly be
described first. But since the one that brings the greatest
profit is much better liked, and justifiably so, I thought
of starting with the most exalted and most precious one,
coming down to the others in a natural sequence. However, although gold because of its nobility easily deserves
precedence, I have deemed it best to describe first the
origin and birth of silver ores because the country around
Meissen (where this little book on the ores was recently
composed), though endowed with all the metallic ores,
abounds in silver ores.

According to the opinion of the philosophers, silver
ores are made through the influence of the Moon (as
mentioned before) from clear quicksilver and expurgated,

familiar with the term. Referring to *Witterung* as an atmospheric phenomenon, Dr. Tom. F. W. Barth, however, wrote:

"Around 1915, when I was a boy, old miners at Röros (the copper deposits
at Röros, Central Norway, were first opened in the 17th century by
German miners) knew about the *Witterung*. One of them pointed out the
phenomenon to me. It was a warm and sunny day, and what he showed
to me was the trembling in the air due to refraction of the light rays in
an ascending hot current of air. The old man was quite aware of the
natural explanation, but told me that this was the *Witterung*, which by
his simple predecessors was taken as an indication of ore. He went on to
explain that where ore crops out, the earth will be warmer than elsewhere
when directly exposed to the sun, therefore the ascending current. Like-
wise he explained that lightning preferably would strike the outcrop of a
vein, because lightning is attracted by metals."

In the illustrations, the *Witterung* is designated *w*.

Silver Ore

strong sulphur, the Moon representing the power of the maker, and quicksilver and sulphur having the proper qualifications of matter. Silver ore occurs in various ways. Some, in river sand, as a black or grey substance similar to what is described later in the chapter on gold ores. Some occurs in veins and stringers, which is discussed in this chapter.

To find a rich and persistent vein, which is to be worked in preference to any other kind, it should be remembered that the most likely location for [12] such a vein is in the south slope of a mountain. If the vein strikes from 7 or 6 east toward 6 or 7 west* (according to the quartering of the earth previously explained), if the outcrop of the whole vein lies north, if its rock outcrop is in the east, if its hanging wall is in the south and its footwall in the north, such characteristics of the mountain and the vein lend themselves readily to receive the influence of Heaven that is to be exerted upon the matter from which silver ore is made or created; and it will also retain the influence securely, like a well-made vessel, so that the creation of silver ores can be brought to perfection. It is thought that veins striking from other points toward the points between west and north, but with the same hanging walls, footwalls, and outcrops are richer or poorer in the measure as they approach or deviate from the aforementioned strike. Veins that strike from north to south, whose hang-

---

*The figures should not be reversed the second time.

Agricola, who in the third chapter of his *De re Metallica* (1556) leaned heavily on the *Bergbüchlein*, has a paragraph very similar to the following. He describes the same vein running from 6 or 7 east to 6 or 7 west—but it is in the north slope of a mountain—and he mentions the same specks and lumps of silver in the less promising ones. He adds, however, that recent discoveries of rich silver veins prove that the theories of the old miners were not founded on fact. He actually quotes von Kalbe (see p. 49) in describing the "favorable orientation" of gold-bearing streams but again brands as superstitious the theory that orientation indicates richness of a deposit and scoffs at the belief that gold originates in streams. He had explained in 1546, in *De Ortu et Causis Subterraneorum*, and so had Biringuccio before him (*De La Pirotechnia*, 1540), that alluvial gold deposits do not originate in streams but are formed by washing down from higher deposits.

Bergbüchlein

ing wall is in the west and whose footwall and outcrop are in the east are more promising to work than veins striking from south to north, whose hanging wall is in the east and whose footwall and outcrop are in the west, although the latter sometimes yield specks of silver attached to the rock and lumps of native silver or a rich ore at several places. But there is nothing persistent and lasting about them, [12v] because all the mineral Power is exhaled and removed and extracted through the outcrop of such veins. The same is true of veins that strike from east to west as the ones above but whose outcrop and footwall are in the south, since their outcrop deprives them completely of the noble substances.

Among the silver veins there are some whose hanging wall and footwall contain quartz; others contain spar, or hornstone, or ironstone. Some contain a white sticky rock, some limestone,* some mixed or speckled rock of many colors (testifying to the mingling of different exhalations since these color the rock), and others contain still other kinds of strange rock.

Within their walls the veins sometimes contain white or yellow pyrites,† or galena, or bismuth ore; sometimes they contain a yellowish or pale earth,‡ or white, brown, or black clay; sometimes, depending on the mineral exhalations, there will be a burnt-looking or a black, blue, brown, or green alteration product;§ and in some cases you will find translucent or dark blende, or a translucent white gangue resembling alum. Some people call the latter transparent quartz although, contrary to the nature of quartz, it is easily liquefied by fire. And what has been said just now about the gangue and minerals in veins

*Reading *kalkstein* for *klagstein*. Dr. Barth pointed out that there is a *klangstein*, phonolite, but that would seem too specific here.

† *Kiese*, i.e., sulphide ores, such as arsenical pyrites, iron and copper pyrites, etc.

‡*Schweif*. According to von Dechen's (see p. 54) annotations, this means any kind of rock that gradually peters out. It is usually soft and clayey.

§*Witterung*. This word seems to be replaced by *Mulm* in more modern literature. See footnote, p. 33, second paragraph.

Silver Ore

is true also for stringers. Wherever these different minerals in veins and stringers contain silver, you should follow them and open them up in the hanging wall and footwall, because such minerals are a sign that points to and leads you to the ore [13] of the steep or the sloping vein.

Wherever branch veins, hanging-wall stringers, or cross veins branch off or occur above the one that determines the strike, that is, the main vein, you may confidently follow it down because veins are enriched at such places and become very rich at depth, provided that of the outcrops of these veins and veinlets one is in the north and the other one in the east. It is advisable, for this reason, to drive off the main vein every so often to look for other connected ore bodies whose outcrop and dip are in favorable directions according to what has been said before, since in some mountains very many smaller ore bodies occur in various directions near a vein.

Wherever in a mountain all kinds of veinlets join when they reach a vein and stay together on their descent into depth, you may confidently start to mine. Especially if silver has been found in a veinlet, the chances are that you will find rich ore at depth. However, if in sinking, fractured rock is struck, do not be discouraged but follow the ore and go all the way through the fractured rock until the vein with its ore is back again in solid rock. And if you strike a barrier, that is, a hard rock or stone that cuts off the vein or the ore but leaves between hanging wall and footwall a clay or some other favorable mineral matter, follow that undismayed.

Where spar veins become weathered and mixed with clay toward the depth, [13v] you will surely be able to mine ore at depth. It is, therefore, very much worthwhile to sink a shaft. Even if one or two of the vein minerals peter out, it is perfectly safe to follow the vein down, if another kind joins in, especially if it is yellow or burnt-looking pyrites that might itself contain silver.

Bergbüchlein

If, in sinking, spar has been mined and the spar is exhausted but no ore occurs soon afterwards, it is to be feared that such spar occurrence was not an ore vein but breccia or gouge, which often run parallel to veins. It should be remembered that all branch veins or hanging-wall stringers which come down from the north or close to it enrich the main vein, which makes it promising to start operations and go down. But hanging-wall stringers coming down from the south or close to it impoverish all veins toward which they point. Similarly, all veins are impoverished by stringers* (whether hanging-wall stringers, intersecting, or cross stringers) or by other ore bodies whose outcrop is in the south or close to it.

Furthermore it should be remembered that sloping veins usually run close to steep veins, and where stringers, cross veins, or other ore lenses and flat veins cross or traverse both the steep and the sloping vein, it is very promising to start mining operations and to sink a shaft, especially if the outcrop of the smaller ore bodies that cut crosswise is in the east and the outcrop of the main vein or the sloping vein is in the north. The prospects [14] are also good if the sloping vein can be connected crosswise with the steep vein approaching it in strike because their union or alliance promises great riches at depth. Thus it is advisable here or there to break into the hanging wall or footwall of the steep vein in order to see if the sloping vein can be reached and to find out by which crosscut and how far away the sloping vein will join the steep vein. Owing to such foresight many a fortune has been built on the union of veins, at very little expense. Since the sloping vein can usually be connected with the steep vein by a crosscut, some mining laws stipulate that the sloping vein shall pay royalties to the steep vein.

To clarify previous remarks about the ore-bearing pos-

*Reading: *die klüffte*, as in the undated edition, instead of *und klüffte*, as in the 1518 edition.

sibilities of veins per se, it might be said that those veins whose hanging and footwalls contain quartz, spar, or hornstone and that have a clayey substance between the walls are good prospects. Those veins that have ironstone or some other iron ore* in their hanging and footwalls but bismuth ore or a fatty tough alteration product between their walls are also promising. Similarly, those veins whose hanging and footwalls are a white country rock and that contain a black or burnt-looking alteration product may be worked profitably if their strike and outcrop conform to what has been said before. Veins that are enclosed and bounded by solid rock [14v] but carry soft or slaty gangue mixed with silver glance or some other barbed, hard ore† between their walls are very promising because such veins may become very rich at depth if they are enriched by other minerals or by having the proper strike, dip, and outcrop as previously discussed.

The Fifth Chapter

On Gold Ore.

GOLD, however, according to the opinion of the philosophers, is made from the very finest sulphur—so thoroughly purified and refined in the earth through the influence of Heaven, especially the Sun, that no fattiness is retained in it that might be consumed or burnt by fire, nor any volatile, watery moisture that might be vaporized by fire—and from the most persistent quicksilver, so perfectly refined that the pure sulphur is not impeded in its influence on it and can thus penetrate and color it from the outside to its very core with its per-

*Eisenstein oder eisenmal.

†Stachel frisch erz. Stachelerz may be the same as Speererz (both meaning barbed ore, or prickled ore) which is an old name for marcasite. T.F.W.B.

sistent shade of citrine. And thus the two, sulphur and quicksilver, being the mineral matter, are joined into a metallic body in the most powerful and enduring union through the influence of Heaven, delegated to the Sun, and through the fitness of the location, through which [15] the mineral exhalations of sulphur and quicksilver wind and drive and break their way. And such union cannot be dissolved even by the most violent and most powerful effort of fire.

Gold occurs in different ways. Some, in ordinary river sand, some under the overburden near swamps, some in pyritic deposits, some, as the native metal, in stringers and veins, and some in various ores and alteration products contained in veins and stringers, whether these are schists, or black, brown, grey, blue, or yellow alteration products, or clayey ores. The gold generated in river sand is the purest and most exalted kind because its matter is most thoroughly refined by the flow and counterflow of the water and also because of the characteristics of the location where such gold is found, that is, the orientation of the river in which such placer gold is made.

The most suitable location for a river is one between mountains in the north and a plain in the south or west. And the most suitable direction of the current is from east to west. The next best is from west to east, with mountains located as described before. The third best is from north to south, with mountains in the east. But the worst, as far as the generation of gold is concerned, is from south to north if high mountains rise in the west. The possible directions [15v] of the flow of water are as manifold according to the quarters of the earth, as those of the strike of veins, which was described earlier in the chapter on silver ores. And each direction is judged better or worse in the measure as it approaches or deviates from what has been said above.

The better to recognize such locations and streams that carry gold, it should be remembered that in general gold

Gold Ore

is likely to be born in streams in which precious stones are found, such as amethysts, rubies, rock crystals, and other highly refined pebbles, which are an indication [16] of the fitness of the place. According to the opinion of Albertus Magnus, hot and dry fumes or exhalations are seldom extracted from the earth without being accom-

[Fig. 12]

panied by warm, moist vapors. The gem stones are wrought and born of dry fumes; and the clearer, finer, and nobler the fumes are, the more beautiful and the better and harder will be the gems. Metals are wrought and made from moist vapors, and how strong and good the metal is will depend on how clear, pure, and well-digested* the matter is from which the vapors or mists are extracted. Since moist and dry exhalations rise together, but each is hardened according to its own nature,

*wolgedewet; deuben is keltern (to press or tread grapes).

Bergbüchlein

it is a very reliable indication of the occurrence of gold, as said before, if precious stones are found in a river. Also, where you find in a river or nearby little crystals of tourmaline of a dense, fine structure a gold occurrence is not far off. It is, however, essential that the crystals be very fine because where the coarse kind is found, there is little hope for an occurrence of the best and finest of the metals, the gold. The value and actual gold content of the gold that is generated under the overburden near swamps depends on how much of the grey or black [magnetite] sand that together with the little leaves or grains of gold constitutes the schlich is mixed up with it in smelting. In many places this schlich contains more silver than gold, and sometimes even copper, so that the gold is less valuable [16v] through admixture with silver.*
This sand may also contain an impurity which darkens the noble and exalted color of the gold so that it gives the impression of being low-grade gold. In reality this subtracts only a little from its value since by some minor deft manipulation such impurity can be removed from the gold to restore its exalted color.

For more information on a likely place for the generation of this kind of gold you should know that gold schlich can probably be coaxed out of spots where many little weathered furrows† are found under the soil where the placer gold occurs. These resemble the little veins or cracks that are sometimes found to run through the loam in loam pits. And where the little furrows bunch or multiply, they multiply and increase the mineral Power from the earth so that more gold is generated.

Gold that is generated in a pyritic deposit is mixed with many and varied worthless impurities because pyrites is made from contaminated sulphur and an im-

*This indicates that despite the availability of three different methods of parting (cementation, sulphur, and acid), the operations were popularly considered to be expensive and undesirable. C.S.S.

†It is now known that these furrows or ridges in bed rock trapped the gold as it washed along the stream bed. S.F.K.

Gold Ore

pure earthy matter. But through the influence of the Sun and Heaven, and given enough time, the finest part of the pyrites is gradually cleansed and boiled into a persistent gold ore, which must be separated from the impure pyrites by the industrious application of strong fire.

[17] Such gold-bearing pyrites is found in some places as bedded deposits that extend through the rock as a complete stratum; according to regional custom, these are sometimes called horizontal veins. Others occur in the form of [fissure] veins, that is, as upright veins that have hanging and footwalls.

The flat-lying pyritic deposits are very low in gold content because the influence of Heaven, owing to the lack of fitness of the position, can exert itself but little.

The gold-bearing pyrites that occur in veins are supposed to increase in richness and gold content in the measure as the country rock of the hanging and footwalls of a vein becomes finer and richer. And depending also on whether or not the strike and outcrop of a vein are in the right direction and whether a vein encounters other stringers that enrich it, as was explained in the chapter on silver ores, the occurrence will vary in quality and gold content. Of the gold that is generated in other than pyritic veins some is found as native gold attached to the rock, some in a yellow clay, some in a brown, fine alteration product, and some finally mixed and worked in with quartz.

Where this brown alteration product occurs as a vein, the prospects are very good; because, with added ore from hanging-wall stringers, it will become very rich at depth.

Similarly, where the yellow clay occurs as a vein, it is promising to mine, provided the vein has a fine country rock in its hanging [17v] and footwalls. Furthermore, where native gold is found in stringers that run near a vein, it should be observed where the stringers join the vein; and there you may confidently start to mine and

Bergbüchlein

sink a shaft. If, however, such stringers swerve away from the vein, you are likely to be disappointed unless they join another vein. Wherever hanging-wall stringers that contain native gold leave a vein either sideward or downward, it is advisable to explore for other veins; by such foresight the stringers and the veins may be worked together.

The Sixth Chapter

On Tin Ore.

TIN ore, or cassiterite,* is generated, through the influence of Jupiter, from pure quicksilver and a little sulphur; and in the union of these two, impure coarse sulphurous vapors are introduced, which are combined into a body and fused into a metal that is called tin. It is because of these impure vapors that all tin is evil-smelling and crackly† and brittle, and that it contaminates all metals with which it is mixed and renders them brittle.

Some cassiterite is born in rivers, as you have heard before of gold, and some of this is washed out as coarse grains resembling tourmaline. [18] This kind, called stream tin, is the nicest and best tin because its matter is thoroughly refined and it is exalted through the characteristics of the location where it occurs. Some tinstone is generated in rocks and found in veins. This kind improves as it occurs farther away from pyritic veins and is less mixed up with pyrites, especially the firm, copper-bearing kind that is very difficult to separate from tin ore. Barren

*Zwitter.

†The cry of tin (caused by the sudden twinning of the crystals under deformation) was supposed to be a sign of deficiency in purity and metallic properties. Actually even the purest tin cries. Tin itself is not brittle, though its alloys with most other metals (except in small amounts) are brittle. C.S.S.

pyrites, on the other hand, is not so very harmful to tinstone because it can be loosened and turned into ashes through the bite of the fire and then be washed off the tin ore with water on a hearth.*

Another kind of cassiterite or tinstone found in rocks occurs in the pockets of a stockwork but not in veins. This kind too is purer and better if it is found farther away from pyritic veins and is less mixed up with the kind of sulphur that is full of iron.

An indication of the occurrence of this tinstone is that it generally effloresces at the outcrop and, in doing so, forms crusts that peel off.

The Seventh Chapter

On Copper Ore.

COPPER ore is generated through the influence of Venus from good, pure quicksilver that, however, has not been delivered entirely of all extraneous excess moisture, and from excessively hot, burning, and impure sulphur. The heat of the sulphur colors the whole metal red through and through.

[18v] This metallic ore is found partly in schistose bedded deposits and partly in veins containing different ores, some of which are brown, some green, and others pyritic. The schistose copper ore is mixed up with a great deal of barren gangue so that it is difficult to obtain the metal by simple roasting or smelting. The kind occurring in veins is better and richer if the hanging and footwalls enclosing the vein are of rich, mineral-bearing country rock. But the quality and richness of the copper ore in veins also depend on whether a vein strikes from the right points of the earth (as was explained before in con-

*Test.

nection with silver veins) and whether it is enriched by intersecting stringers and veinlets.

What you have heard before of the enrichment of silver veins is equally true of the strike and enrichment of copper veins. Only, those copper veins that lie in the north slope of a mountain are usually wide, but their copper is relatively poor in silver, while those that lie in the south slope of a mountain are relatively thin, but their copper contains more silver. And these veins too are enriched by the direction of their strike, as was discussed previously in connection with silver veins.

---

[19] ## The Eighth Chapter

On Iron Ore.

IRON is generated through the influence of Mars from impure quicksilver and brittle, impure sulphur that in fusing into the metal introduces into it a great deal of earth, which makes it very difficult to soften iron by fire. It also contains very much rust on account of the impurity of the sulphur, which is the reason why it cannot easily be mixed, cast, or alloyed with another metal.

Iron ore is found at some places in a group of irregular layers of brown or yellow color; at some others, in veins. Iron ore of the first kind yields a great deal of scoria or iron slag and little iron. Although the iron ore occurring in veins is richer in iron, the latter is sometimes brittle because it is mixed up with some other kind of metal.

But iron veins well bounded by hanging and footwalls should not be despised, provided their strike is from east to west and their dip toward the south. And if their footwall and outcrop are in the north, you will generally find gold or some other valuable ore after the iron ore has been mined.*

*Rich veins often have iron ore at their outcrop—the iron cap.

Other Ores

The Ninth Chapter

On Lead Ore.

LEAD ore is generated through the influence of Saturn from unrefined, watery, heavy, dirty quicksilver and a little sulphur that, through its expanding [19v] hot vapors, boils and coagulates the quicksilver into a metallic body. And because the two, sulphur and quicksilver, are only loosely bound in this union, the resulting metallic body, the lead, is easily consumed and driven off by fire.

Some of the ore of this metal is found in flat veins and some in steep veins. Lead ore from flat veins under the overburden is low in silver, unless silver-bearing minerals are fed into it, which may be done by stringers. The lead ore that occurs in veins varies in richness and silver content depending on the strike, dip, and enrichment of the vein as it was explained previously in connection with silver. Some of the lead ore occurring in veins is black, some is minium-colored, and some is lustrous.*

The Tenth Chapter

On Common Quicksilver.

COMMON quicksilver† is generated through the influence of Mercury from muddy, watery moisture mixed with the very finest sulphurous earth. This metal is found sometimes in a brown earth, just as any other metallic ore, sometimes in pits into which it has seeped out of the fissures of the earth as water seeps into swamps. Some is exhaled from the earth and after drifting over the earth's surface, precipitates on the grass. [20] This metal is of a strange nature, as the alchemists have good

*glanzig; reference to lead glance, galena.
†As opposed to elemental quicksilver.

reason to know. And I shall leave it to them to continue their squabbling about it.

Knappius: Even if I know now about the matter from which ores are made and about the locations where they are likely to be generated, this does not tell me how one or the other of them might be smelted into its metal.

Daniel: The shift is up, enough for today. Tomorrow we shall go from the mine shack to the smelter, and I shall tell you which fluxes to use on sulphur-bearing ores, which on easily fusible ones, which on contaminated ones, and which on coarse- or fine-grained ones.

If somebody, in order to make the mountains stand out more clearly and plainly, should like to have the figures brushed or painted, the veins might be shown yellow, the mist and shimmering smoke-colored, and the water blue. To indicate which is which, I have severally used the following lettering:

g veins (*gänge*)
w shimmering (indicating ore) (*witterung*)
n mist (*nebel*)

[This is followed by eight pages of a glossary of mining and smelting terms; under this, on p. 24:]

Printed at Worms at Peter
Schöfer's and finished on
the fifth day of April
M. D. XVIII.

History and Bibliography of the Bergbüchlein

WHEN Georgius Agricola, in 1550, dedicated his now famous *De re Metallica* to the rulers of Saxony, Thuringia, Meissen, and other, smaller principalities, he included in his review of previous works on similar subjects reference to two books written in German: one, an "anonymous, rather disorganized account of assaying for metallic content and of metals," which must be the *Probierbüchlein*, the other, "dealing with veins, written by Calbus of Freiberg, a well-known physician."* Calbus is mentioned a second time—and cited—at the end of the Third Book of *De re Metallica*. Comparison of the citation with the *Bergbüchlein* leaves no doubt that Agricola considered Calbus of Freiberg the author of the latter work, although all the editions of the *Bergbüchlein* known now were published anonymously.

It is generally accepted by those who have dealt with the bibliography of the *Bergbüchlein*† that Calbus was Ulrich Rülein von Kalbe, who served the City of Freiberg, in Meissen, as health officer in 1497, as member of the City Council in 1509, and as mayor in 1514 and 1519. He died at Leipzig in 1523. In connection with the very garbled description of the mining compass toward the end of Chapter 3 of the *Bergbüchlein* it is of interest that von Kalbe is supposed to have been not only a physician of repute but a mathematician, astronomer, mine sur-

---

*The Latin reads: Nostra quidem lingua duo libri scripti sunt: alter De materiae metallicae et metallorum experimento, admodum confusus, cuius operis parens ignoratur: alter De uenis, de quibus etiam Pandulfus Anglus scripsisse fertur: sed librum Germanicum confecit Calbus Fribergius, non ignobilis medicus.

†See, for instance, Herbert C. and Lou H. Hoover, translation into English of *De re Metallica*, London, 1912, Appendix B.

Ernst Darmstaedter: *Berg-, Probir- und Kunstbüchlein*, Munich, 1926.

F. D. Adams: *Birth and Development of the Geological Sciences*, Baltimore, 1938.

Bergbüchlein

veyor, and architect, which raises some doubt whether this description actually originated with von Kalbe. It might have been mutilated in printing; but its position in the book makes it seem more likely that it was added later.

Compared with the *Probierbüchlein*, the story of the various printings (see pp. 55 to 62) of the *Bergbüchlein* is simple. It is generally believed that of the oldest edition known, only two copies are still in existence: one at the Bibliothèque Nationale of Paris, the other at the Staats- und Stadtbibliothek of Augsburg. Both are now undated and show no place of publication. The French list their copy among the incunabula,* but Darmstaedter concluded from inspection of the copy at Augsburg that this edition of the *Bergbüchlein* was printed between 1505 and 1510 "probably" by Martin Landsberg at Leipzig.†

Of the later editions of the *Bergbüchlein* the one printed at Erfurt in 1527 resembles the undated edition most closely. There are some additional illustrations, but they serve only for decoration; i.e., there is in the 1527 edition in each of the later chapters dealing with the different metallic ores a picture of the sign of the zodiac and also of the deity associated with the planet held to be influential in the generation of the ore under discussion.

There is an earlier reprint—or one might call this a second edition—published by Peter Schöfer at Worms in 1518. It is this edition which was reissued repeatedly: by Heinrich Steyner of Augsburg in 1534 and 1539; by Christian Egenolph of Frankfurt a.M. as the *Bergwerck und Probirbüchlin* in 1533 and 1535; and by an unknown

*Marie Pellechet: Catalogue Général des Incunables des Bibliothèques Publiques de France, A. Picard et Fils, Paris; vol. 1, 1897, p. 501.
†Comparison of a microfilm of the Paris copy with the passages of the Augsburg copy that were reproduced by Darmstaedter showed some differences in spelling. Dr. Paul Geissler, Director of the Augsburg library, was kind enough to ascertain that these differences are misprints in Darmstaedter's book, and that the two copies are actually identical. The Augsburg copy, however, is incomplete: two leaves are missing in Chapter 3. Some of its woodcuts are colored.

48

History

printer, on behalf of the publisher Johann Haselberg of Reichenau, in combination with some old mining laws, as *Der Vrsprung gemeynner Berckrecht*, possibly in the 1520's.

In their Agricola translation, the Hoovers state that, according to the librarian of the Bergakademie Freiberg, the collection there includes a 1512 edition (date written in by hand), also by Peter Schöfer of Worms, and that there is a record of a 1512 issue of the *Bergbüchlein* in *Panzer's Annalen*, Nürnberg, 1782, p. 422. This listing in *Panzer's Annalen* was known to an earlier bibliographer of the *Bergbüchlein*, Hans von Dechen,[*] who believed the 1512 to be a misprint for 1518. Darmstaedter, though familiar with the Hoovers' discussion and with the Freiberg collection, makes no mention of a 1512 edition. In answer to an inquiry, Professor G. Gruess, the present head of the Bergakademie's library, wrote on June 3, 1948 that he has no record of the existence of such an edition and that he is satisfied that his library's copy (date written in by hand) is one of Peter Schöfer's 1518 edition.

The difference between the undated and the 1518 edition is only superficial. The publisher of the latter dropped a brief introduction before Daniel and the young miner start their dialogue; he changed the title page and its woodcut; he had the illustrations redrawn and added brief instructions on how to color them; and he appended to the text six pages of a glossary of mining terms—which, however, deal with aspects of mining different from those dealt with in the book—and two pages of metallurgical terms. The text is the same in both editions in spite of some minor omissions in the 1518 issue. Most of these are editorial deletions to reduce redundance, while a few are oversights. There are the expected differences in spelling; but some variations as, for instance, that the undated edition has the unintelligible

[*] *Zeitschrift für Bergrecht*, v. 26, 1885, pp. 219–261.

Bergbüchlein

steinender gang where the 1518 edition has correctly
stehender gang, and the fact that the 1518 edition has
feystigkeyt, while the earlier edition erroneously uses
festigkeit where the context obviously calls for "fattiness"
not "strength," would indicate that the 1518 edition was
copied from an even earlier and more correct version than
the now undated edition. The confusion in the illustrations rules out the possibility of intelligent editing (see
footnote, pp. 25 to 26.)

Of the reissues of the 1518 edition only the 1534 and
the 1539 printings by Steyner of Augsburg were not combined with another work. The *Bergwerck und Probirbüchlin*, printed by Egenolph of Frankfurt a.M. in 1533 and
1535, contains the entire *Bergbüchlein* (without the glossary), short excerpts from the *Probierbüchlein*, and some
miscellaneous material on dissolving metals, polishing
gems, and metal poisoning, some of which can be traced
to the *Kunstbüchlein*; the part on metal poisoning is a reprint of Dr. Ulrich Ellenbog's treatise *Von den gifftigen
Besen Tempffen und Reuchen* written in 1473.

Both F. D. Adams and Hans von Dechen mention the
inclusion of the *Bergbüchlein* in an undated work called
Der Vrsprung gemeynner Berckrecht, which they date
tentatively 1519–1520 and about 1532, respectively. In
this edition, the dialogue has been omitted from the
Bergbüchlein, but the glossary is included. The first half
of *Der Vrsprung gemeynner Berckrecht* consists of very
old mining laws of Bohemia, Moravia, Meissen, etc.,
which are followed by a new caption: "*Von erkantnus der
Berckwerck.*" The first page of text under this caption
might be called an abstract of the omitted dialogue of the
Bergbüchlein and serves to introduce the ten chapters of
the original text. These are followed by the *Bergnamen*
and finally by three pages listing the ores and mines of
Bohemia.

Der Vrsprung gemeynner Berckrecht shows no place of
publication and has no page numbers, but it is dedicated

History

by Johann Haselberg of Reichenau, a book publisher, to Johann Lucas, in gratitude for favors shown him by Lucas while the latter was counselor to the "late Emperor Maximilian." This proves that the book was printed after January 12th, 1519, when Maximilian died, and before 1538, when Haselberg's publishing business failed.*

Dr. Adams accepted the date of 1519–1520, and the publisher as Knobloch (i.e., Johann Knoblauch) of Strassburg, on the authority of a German book dealer's statement that the type used indicates Knoblauch, who, according to F. Kapp's *Geschichte des Deutschen Buchhandels*, printed for Haselberg "around that time." However, Kapp states only that Haselberg employed Knoblauch in 1516 and, according to Roth,* Haselberg's activities as a publisher between 1515 and 1538 extended over a very large part of Germany.

Hans von Dechen based his estimate of 1532 on the fact that the copy of the book which he inspected was bound together with two other small volumes containing old mining laws, which were printed that year by Friedrich Peypus of Nürnberg and for which "apparently" the same type and style were used. (Peypus was active as a printer from 1509–1535.)

Comparison of two pages of *Der Vrsprung gemeynner Berckrecht* as reproduced by von Dechen with photostats of the same pages of Dr. Adams' copy shows them to be identical. It is safe, therefore, to conclude that both refer to the same edition, another copy of which is in the Parsons Collection of the Reference Department of The New York Public Library, tentatively dated 1530 there.

It is a sign of the historical interest of the *Bergbüchlein* that it was reprinted at least once in each of the next three centuries:

In 1698, from the Haselberg edition, by Johann David Zunner at Frankfurt a.M. as part II of a work entitled

*F. W. E. Roth, *Archiv für Geschichte des Deutschen Buchhandels*, v. 18, 1896, pp. 16–28.

Bergbüchlein

Corpus Juris et Systema rerum Metallicarum, oder: Neuverfasstes Berg-Buch. In this reprint, the woodcuts are replaced by copper plates. (Engineering Societies Library, New York.)

In 1792 by J. F. Lempe, from the 1534 edition, in Part 9 of *Magazin der Bergbaukunde* (pp. 21–56), without comments and without illustrations. (Engineering Societies Library, New York.)

In 1885, by Hans von Dechen, from the 1539 edition, in *Zeitschrift für Bergrecht*, v. 26, 1885, pp. 219–262, as part of an article: "*Das älteste deutsche Bergwerksbuch*," which contains technical annotations and an extensive bibliographic discussion. Parts of this discussion are influenced by the fact that von Dechen was unable to locate a copy of the undated edition and did, therefore, not realize that the 1527 edition is, in some ways, older than that of 1518. (Law Library, Columbia University, New York.)

List of Known Editions of the *Bergbüchlein*

IN listing the known editions of the *Bergbüchlein* it seemed preferable to reproduce rather than describe the title pages, and to add to these reproductions brief descriptions of the colophon and other bibliographical notes. Information on those copies which were not available either in the original or as photocopies is based on Darmstaedter and von Dechen. It seemed desirable to indicate which editions are represented in the libraries of the United States and Canada; but no attempt was made to trace or list all available copies.

Edition A: *Undated* (no place of publication)

According to Darmstaedter, printed between 1505 and 1510, probably by Martin Landsberg at Leipzig.

24 (unnumbered) leaves. Copies at Bibliothèque Nationale, Paris, and at Staats- und Stadtbibliothek, Augsburg. Microfilm at University of Chicago Library.

Eyn wolgeoꝛdent vnd nütz-

lich büchlin / wie man Bergwerck suchen vñ
finden sol, von allerley Metall, mit seinen figuren,
nach gelegenheyt deß gebirgs artlich angezeygt. Mit anhangenden Berck'na-
men den anfahenden bergleut-
ten vast dinstlich.

Edition B: 1518 (Worms)

Colophon: Getruckt zu Wormbs bei Peter Schöfern/vñ volen-
det am fünfften tag Aprill. M.D.XVIII.

24 (unnumbered) leaves, including 8 pages of *Bergnamen*.
Copy at Montreal, McGill University; photocopy in translator's possession.

Ein nützlich Berg

büchlin von allen Metal-

len/als Golt/Silber/Zcyn/Kupfer
ertz/Eisen stein/Bleyertz/vnd
vom Queckfilber.

Edition C: 1527 (Erfurt)

Colophon: Gedruckt zu Erffurd/durch Johan Loersfelt. 1527.
24 (unnumbered) leaves. Copy in Hoover collection.

Der Vrsprung gemeynner

Berckrecht/wie die lange zeit von den alten er-
halten worde/darauß die Küniglichen vñ Fürstlichen bergks ord
nungen vber alle Bergre.be geflossen/ welcher sich eyn ieg-
licher in zufelligen Berckhandlungen/ vor dem obristen
Berckmeister vnd anderen Berckrichtern/ zu recht
wol gebrauchen mag/Auch ein anzeygung der
clüffe vnd geng des Metallischen Artz/wie
die in berg vnnd thal streichent/ vnd
ibt geschick haben/ Mit art-
lichen figuren ver-
zeichnet.

Sampt eyner anzeygung vil böslicher vnd sündiger
Berckwerck der löblichen Cron zu Beham.

Edition D: *Undated* (no place of publication)
Der Vrsprung gemeynner Berckrecht

Colophon: Durch Johan Haselberger auss der Reichenaw/in druck verordnet.

44 (unnumbered) leaves. The *Bergbücblein* proper occupies the second half of the book. Copies at The New York Public Library and at McGill University, Montreal.

(The title-page reproduction was made from the copy at New York, which is mutilated at the bottom. Other copies have under the woodcut a line reading: CVM GRATIA ET PRI-VILEGIO C. M.)

Bergwerck

vnd Probir büchlin. Für die Bergk vnd Feurwer-
cker/ Goltschmid/ Alchimisten vnd Bünstner.

¶ Gilbertus Cardinal vonn Soluiren vnnd
scheydungen aller Metal.

¶ Polirung allerhand Edelgesteyn.

¶ Fürtrefliche Wasser zum Etzen/ Scheyden
vnd Soluiren.

¶ Verhütung vnd rath für gifftige dämpffe
der Metal.

Edition E: 1533 (Frankfurt a.M.)
Bergwerck und Probirbüchlin

Colophon: Zu Franckfurt am Meyn/ bei Christiau Egenolph.
Im Herbstmon/ Des Jars M.D.XXXIII.

39 leaves, including 2 pages of index. The *Bergbüchlein*
proper occupies leaves 10 to 26. Copy in Hoover collection.

Ein wolgeordent vñ

nutzlich Büchlein/ wie man Bergwerck
suchen vnd finden sol/ von allerley Metall/ mit
seinen figuren / nach gelegenheyt deß ge-
bürges / artlich angezeygt/ Mit anhangenden Bercknamen / den
anfahenden Bergleüten
vast dienstlich.

M. D. XXXIIII.

Edition F: 1534 (Augsburg)

Colophon: Getruckt zu Augspurg durch Heinrich Steyner/Am
3. tag Octobris/Im M.D.XXXIIII.

55 pages, including 10 pages of *Bergnamen.* Copy at library
of Bergakademie Freiberg. Photocopy in translator's possession. Reprinted in *Magazin für Bergbaukunde*, Part 9, 1792,
pp. 21–56. (Engineering Societies Library, New York.)

Bergwerck vñ

Probir büchlin/für die Bergk vnnd fewrwercker
Golschmid/Alchimisten vnd Künstner.

Gilbertus Cardinal vonn Soluiren vnd schei
dungen aller Metal

Polirung aller hand Edelgestein.

Fürtreffliche Wasser zum Etzen/ Scheyden/
vnd Soluiren.

Verhütung vnd Rath für gifftige dämpffe
der Metal.

Zu Franckfurt/bei Christian Egenolff.

Edition G: 1535 (Frankfurt a.M.)
Bergwerck und Probirbüchlin

Colophon: Zu Francfurt am Meyn/bei Christian Egenolph
An. M.D.XXXV.

36 leaves, including 2 pages of index. The *Bergbüchlein*
proper occupies leaves 10 to 24. Copy at Montreal, McGill
University. Photocopy in translator's possession.

Ein wolgeordent vñ
nutzlich büchlin wie man Bergwerck
suchen vnd finden sol / von allerley Metall / mit
seinen figuren / nach gelegenheyt, deß ge
birges / artlych angezeygt / Mit anhangenden Bercknamen / den
anfahenden Berglenten
vast dienstlich.

M.D.XXXIX.

Edition H: 1539 (Augsburg)

Colophon: Geruckt zu Augspurg durch Heinrich Steyner/Am
j. tag Augusti/Im M.D.XXXIX. Jar.
Reprinted in *Zeitschrift für Bergrecht*, v. 26, 1885, pp. 222–
257, from which this cut was reproduced. It is obvious that
the text above the original woodcut was reset in 1885. (New
York, Columbia University, Law Library.)

Notes on Translation and Acknowledgments

EVERY English-speaking mining engineer or geologist knows that some terms used by his profession are not rigid but vary regionally; for example: what is called a vein by some, is called a lode by others and is known as a reef to their Australian colleagues. There are parallels in the German language. Since, in addition, the meaning of some words may change considerably in 450 years—and other words may disappear completely—and not only the terminology but the concepts of a science are subject to drastic changes over such a long period of time, translating the *Bergbüchlein* posed some interesting problems. Old German texts and glossaries such as C. T. Delius' *Anleitung zu der Bergbaukunst* (1773), Minerophilus' *Neues und Curieuses Bergwercks-Lexicon* (1730), and H. Veith's *Deutsches Bergwörterbuch* (1871) proved useful sources of information when definitions of obscure terms were needed.

The task of translating was made even more challenging by the author's personality. As Ulrich Rülein von Kalbe was primarily a physician, his information must have been mostly secondhand. Since the nomenclature of the miners of Saxony, where he gathered his material on silver ores, undoubtedly varied from that of the German miners in Hungary, for instance, who probably supplied him with data on gold ores, it is not astonishing that he is not consistent in his use of some expressions. To give an example: there is every indication that *Flöz* and *schwebender Gang* (see footnote p. 21) are used interchangeably in spite of his occasional disapproval of such usage.

It is obvious that he was an educated man with a questing and active mind and, although he apologizes—not wholly sincerely, one might suspect—for the simple

Bergbüchlein

language in which his little book is written, his interest in words, their derivation, and especially their multiplicity of meaning is evident throughout. This gives the original unusual charm but can, unfortunately, not be reproduced in a translation. If the word *Gebirge*, for instance, is used in four ways (as mountains, terrain, rock, and gangue); if *Geschick* can be a very small vein, mineral matter, ore, and a number of other things; if the author uses *Geschicklichkeit* sometimes in the sense of *Beschaffenheit*, i.e., nature—but often with the implied meaning of nature favorable to produce ore (*Geschicke*)—and sometimes to mean ore, or ore shoots, this play on words makes delightful reading but the interpretation of some passages by different translators might conceivably not be the same.

To clear up some uncertainties, authoritative advice was sought. It is pleasant to acknowledge indebtedness to Mr. Sherwin F. Kelly, Dr. Joseph T. Singewald, and Dr. Robert Sosman, who read the first draft of the manuscript and commented on different parts; to Mr. Edward H. Robie and Dr. Cyril Stanley Smith, who generously gave their time to criticize and argue; to Professor Chas. H. Behre, Jr. for some helpful comments; and, especially, to Dr. Tom. F. W. Barth, who was kind enough to compare the revised draft with the German text, suggest some changes, and spend a day in discussing some of the controversial points. Annotations by those consulted are initialed.

Dr. W. W. Francis, of the Osler Library of McGill University, was kind enough to give me the needed information on Dr. Adams' copy of *Der Vrsprung gemeynner Berckrecht* and to help me obtain microfilms and photostats. And finally, the excellent collection of rare books on mining and related subjects at the Engineering Societies Library at New York proved invaluable throughout.

A. G. S.

Three Figures from the Early Undated Edition

[Fig. 11]

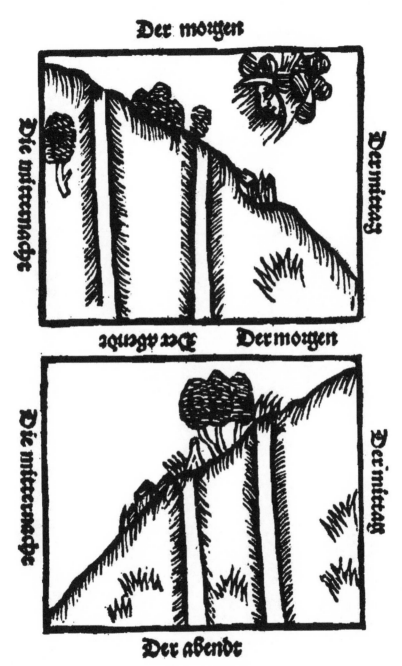

[Figs. 3 and 4]

INDEX

Numbers refer to page numbers. The letter "n" following an entry refers to a note on that page.

CPSIA information can be obtained
at www.ICGtesting.com
Printed in the USA
LVHW030755040720
659720LV00002B/743